Mathematical Formulas for Economics and Business:

A Simple Introduction

Also by K.H. Erickson

Simple Introductions

Accounting and Finance Formulas
Choice Theory
Corporate Finance Formulas
Econometrics
Financial Economics
Game Theory
Game Theory for Business
Investment Appraisal
Mathematical Formulas for Economics and Business
Microeconomics

Mathematical Formulas for Economics and Business:

A Simple Introduction

K.H. Erickson

© 2014 K.H. Erickson

All rights reserved.
No part of this publication may be reproduced, stored in or introduced into a retrieval system, or transmitted in any form or by any means, including electronic, mechanical, photocopying, recording or otherwise, without the prior permission of the author.

Contents

Supply and Demand	6
Market Equilibrium	10
Non-Linear Functions	14
Financial Mathematics	16
Differentiation	19
Functions of Several Variables	23
Integration	27
Matrix Algebra	29

Supply and Demand

A straight line

A straight line can give the relationship between two variables, defined by vertical intercept (a) and slope (b):

Straight line: y = a + bx

The slope is the change in y given a one unit change in x:

Slope: b = Δy / Δx

Δ = Change from one value (x_1 or y_1) to another (x_2 or y_2):

Slope: b = ($y_2 - y_1$) / ($x_2 - x_1$)

Price and quantity

The properties of a straight line may be applied to the relationship between a product's price and its quantity.

Demand function: P = a − bQ

There is a negative relationship between a product's price (P) per unit and the quantity of units demanded (Q).

Supply function: P = c + dQ

There is a positive relationship between a product's price per unit and quantity supplied. Letters c and d represent the vertical intercept or constant, and slope respectively.

With a tax (t) per unit the supply function becomes:

Supply function with tax: P − t = c + dQ

Revenue and costs

Multiplying the price per unit (P) by the quantity of goods sold (Q) gives the total revenue (TR) for a firm:

Total revenue: TR = P x Q

A firm's profit is total revenue minus total costs (TC):

Total costs: TC = FC + VC = FC + kQ

Where FC = Fixed costs;
VC = Variable costs;
k = Cost of production per unit.

With a tax on a product's unit price the total cost becomes:

Total cost with tax: TC = FC + (k + t)Q

Break-even point: TR = TC

The cost constraint (C) combines labour and capital costs:

Cost constraint: C = wL + rK

L = Labour input quantity;
w = Wage rate of labour;
K = Capital input quantity;
r = Rental rate of capital.

The budget constraint for a firm shows how far monetary income (M) can go:

Budget constraint: M = xP$_x$ + yP$_y$

x, y = Number of units of two different goods;
P$_x$, P$_y$ = Price per unit of good x and price per unit of good y respectively.

Elasticities

The price elasticity of demand (ε_d) shows the effect of a change in a product's unit price on the quantity demanded:

ε_d = % Δ **in quantity demanded / %** Δ **in unit price**
= **(ΔQ/ΔP) x (P/Q)**

The price elasticity of supply (ε_s) replaces the quantity demanded in the above formula with the quantity supplied.

The point elasticity of demand gives the ε_d at a single point for the demand function, P = a – bQ:

Point elasticity: ε_d = (–1/b) x (P/Q)

The arc elasticity of demand gives the ε_d between two points, where the first point has price P_1 and quantity Q_1, and the second point has price P_2 and quantity Q_2:

Arc elasticity: ε_d = (ΔQ/ΔP) x [(P_1 + P_2)/(Q_1 + Q_2)]

The income elasticity (ε_y) replaces price and gives the effect of a change in income (Y) on quantity demanded:

Income elasticity: ε_y = (ΔQ/ΔY) x [(Y_1 + Y_2)/(Q_1 + Q_2)]

Market Equilibrium

Goods market

Goods market equilibrium: $Q_d = Q_s$ and $P_d = P_s$

Q_d = Quantity demand;
Q_s = Quantity supply;
P_d = Price demand;
P_s = Price supply.

The equilibrium for goods x and y when they are substitutes or complements can be found with:

Substitute goods: $Q_x = a - bP_x + dP_y$

Complement goods: $Q_x = a - bP_x - dP_y$

Labour market

Labour market equilibrium: $L_d = L_s$ and $w_d = w_s$

L_d = Labour demand;
L_s = Labour supply;
w_d = Wage demand;
w_s = Wage supply.

Labour demand: $w_d = a - bL$

There is a negative relationship between the wage rate (price per unit) and number of labour units demanded.

Labour supply: $w_s = c + dL$

There is a positive relationship between the wage rate and number of labour units supplied.

National income model

Equilibrium occurs when income (Y) = expenditure (E). Expenditure consists of:

(i) Consumption: $C = C_0 + bY$, less tax: $T = tY$

Where C_0 = Constant consumption;
bY = Consumption which varies with income;
t = Tax rate.

(ii) Investment: I_0
(iii) Government expenditure: G_0
(iv) Exports: X_0
(v) Imports: $M = M_0 + mY$

Where M_0 = Constant imports;
mY = Imports which vary with income.

$E = C + I + G + X - M$

Equilibrium exists with a Y_e income level:

Equilibrium consumption: $C_e = C_0 + bY_e$
Equilibrium taxation: $T_e = tY_e$

IS/LM model

IS/LM stands for Investment Saving / Liquidity Preference Money Supply.

The goods market equilibrium occurs when $Y = E$ (national income), and if investment (I) is a function of the interest rate (r):

$I = I_0 - dr$

Then the following equation will hold:

$$Y = (1 / [(1 - b) \times (1 - t)]) \times (C_0 + I_0 - dr + G_0)$$

This means r is a function of Y to give the IS schedule:

$$r = f(Y)$$

The money market equilibrium occurs where money supply = money demand.

Money demand: $M_d = kY + (a - hr)$

Money demand has a positive relationship with Y, income (kY), and a negative one with r, the interest rate (a − hr).

Money supply: $M_s = M_0$

Money market equilibrium: $M_s = M_d, M_0 = kY + a - hr$

This may be denoted differently to give the LM schedule:

$$r = g(Y)$$

Goods and money markets are in simultaneous equilibrium with the values of r and Y which satisfy simultaneous IS and LM equations.

Non-Linear Functions

Quadratic equations

A quadratic equation has the form:

$$ax^2 + bx + c = 0$$

The value of x can be found with the formula:

$$x = [-(b) \pm \sqrt{(b^2 - 4ac)}] / 2a$$

Exponential indices

$$a^m \times a^n = a^{m+n}$$

$$a^m / a^n = a^{m-n}$$

$$(a^m)^k = a^{m \times k}$$

Logarithms

To convert from index to log form, the base of the index becomes the base of the log, and the power drops down:

Index form: Number = base^{power}

Log form: Log$_{base}$ (Number) = power

To convert from log form to index form, the same process occurs but in the opposite direction.

Log rules

$\log_b M + \log_b N \leftrightarrow \log_b MN$
$\log_b M - \log_b N \leftrightarrow \log_b (M/N)$
$\log_b (M^z) \leftrightarrow z \log_b (M)$
$\log_b (N) \leftrightarrow \log_x (N) / \log_x (b)$

Financial Mathematics

Arithmetic sequences

a = First term in the arithmetic sequence;
d = Common difference.
e.g. if a is 1 and d is 3, the sequence = 1, 4, 7, 10, 13,...

nth term: $T_n = [a + (n - 1)d]$

Sum of n terms: $S_n = \{(n/2) [2a + (n - 1)d]\}$

Geometric sequences

a = First term in the geometric sequence;
r = Common ratio.
e.g. if a is 1 and r is 4, the sequence = 1, 4, 16, 64, 256,...

nth term: $T_n = ar^{n-1}$

Sum of n terms: $S_n = a(1 - r^n) / 1 - r$

Future value

Simple interest: $P_t = P_0(1 + it)$

Where P_0 = Current price;
P_t = Price in t years;
t = No. of years into the future;
i = Interest rate.

Annual compound interest: $P_t = P_0(1 + i)^t$

Compounded m times annually: $P_t = P_0[1 + (i/m)]^{mt}$

Continuous compounding: $P_t = P_0 e^{it}$

Present value

Simple discounting: $P_0 = P_t / (1 + it) = P_t(1 + it)^{-1}$

Compound discounting: $P_0 = P_t / (1 + i)^t = P_t(1 + i)^{-t}$

Continuous discounting: $P_0 = P_t e^{-it}$

Annual percentage rate

With the nominal rate (i) compounded m times per year:

APR = [1 + (i/m)]^m – 1

With the nominal rate compounded continuously:

APR = e^i – 1

Differentiation

Derivatives of standard functions

Function → Derivative:

$y = f(x)$ → dy/dx
x^n → nx^{n-1}
e^x → e^x
$\ln(x)$ → $1/x$
K → 0
Kx^n → $K(nx^{n-1})$
$K_1 x^n + K_2 x^m$ → $K_1(nx^{n-1}) + K_2(mx^{m-1})$

Chain rule

$y = f(g(x))$. With $u = g(x)$ and $y = f(u)$:

Step 1: Find du/dx and dy/du

Step 2: Multiply results: (dy/du) (du/dx) = dy/dx

Product rule

With 2 functions of x, u(x) and v(x), y = u(x)v(x) = uv:

dy/dx = v(du/dx) + u(dv/dx)

Quotient rule

With y = u(x) / v(x) = u/v then:

dy/dx = [v(du/dx) − u(dv/dx)] / v²

Marginal functions

A marginal function is the derivative of the total function:

Marginal cost: MC = d(TC) / d(Q)

Marginal revenue: MR = d(TR) / d(Q)

Marginal product of labour: MPL = d(Q) / d(L)

Where TC = Total costs;
TR = Total revenue;
Q = Quantity;
L = Labour.

Average functions

Average cost: AC = TC / Q

Average revenue: AR = TR / Q

Average product of labour: APL = Q / L

Optimization

Maximum profit occurs when the following two conditions are met:

MR = MC

d(MR) / dQ < d(MC) / dQ

Production and labour

Short-run production function: $Q = f(L)$

Total labour costs: $TLC = wL$

Average labour costs: $ALC = TLC / L$

Marginal labour costs: $MLC = d(TLC) / dL$

Where w = wage rate.

Functions of Several Variables

Function of one variable

A function of one variable: y = f(x)

dy/dx = Derivative of y with respect to x;
dy = Differential of y, and infinitesimally small;
dx = Differential of x, and infinitesimally small.

Function of two variables

A function of two variables: z = f(x, y)

First-order partial derivatives:

Partial derivative of z with respect to x, constant y: $\partial z/\partial x$ **(z_x for short)**

Partial derivative of z with respect to y, constant x: $\partial z/\partial y$ **(z_y for short)**

Second-order partial derivatives:

$\partial^2 z/\partial x^2$ (z_{xx} for short)

$\partial^2 z/\partial y^2$ (z_{yy} for short)

$\partial^2 z/\partial y \partial x$ (z_{xy} for short)

$\partial^2 z/\partial x \partial y$ (z_{yx} for short)

Total differential of z:

dz = (∂ f/∂ x)dx + = (∂ f/∂ y)dy

For small (i.e. incremental) changes dx becomes Δx etc., and the formula is:

$\Delta z \simeq (\partial f/\partial x) \Delta x + = (\partial f/\partial y) \Delta y$

Unconstrained optimization

Step 1: Find the first and second order derivatives, z_x, z_y, z_{xx}, z_{yy}, and z_{xy} (which equals z_{yx})

Step 2: Solve first-order conditions, $z_x = 0$ and $z_y = 0$ for x and y coordinates of the turning point. Finding z may also be required

Step 3: Examine second-order conditions.
The point is a minimum if: $z_{xx} > 0$, $z_{yy} > 0$, and $\Delta > 0$;
The point is a maximum if: $z_{xx} < 0$, $z_{yy} < 0$, and $\Delta > 0$;
The point is a point of inflection if the second derivatives have the same sign and $\Delta < 0$;
The point is a saddle point if the second derivatives have different signs and $\Delta < 0$.

Where $\Delta = (z_{xx})(z_{yy}) - (z_{yx})^2$

Constrained optimization: Lagrange multiplier

Given a function to be optimized, $z = f(x, y)$, which is subject to a budget constraint, $M - ax - by = 0$, the Lagrangian (L) is defined as:

$$L = f(x, y) + \lambda(M - ax - by)$$

The values of x, y and λ which optimize L are found with:

$\partial L / \partial x = 0$

$\partial L/\partial y = 0$

$\partial L/\partial \lambda = 0$

National income model multipliers

Investment multiplier:
$\partial Y_e/\partial I = 1 / [(1 - b)(1 - t)]$

Government expenditure multiplier:
$\partial Y_e/\partial G = 1 / [(1 - b)(1 - t)]$

Income tax rate multiplier:
$\partial Y_e/\partial t = (-bY_e) / [(1 - b)(1 - t)]$

Where Y_e = Equilibrium income;
I = Investment;
G = Government expenditure;
t = Tax rate;
b = Slope.

Integration

Rules for integration

Standard function → Integral:

$f(x)$ → $F(x) = \int f(x)\, dx$
K → $Kx + c$
x^n → $[x^{n+1} / (n+1)] + c$
x^{-1} or $1/x$ → $\ln(x) + c$
e^x → $e^x + c$

Where c = Constant term;
$n \neq -1$.

Functions of linear functions → Integral:

$f(mx + c)$ → $F(mx + c) = \int f(mx + c)\, dx$
$(mx + c)^n$ → $(1/m)\,[(mx + c)^{n+1} / (n+1)] + c$
$1 / (mx + c)$ → $[(1/m) \ln(mx + c)] + c$
$e^{(mx + c)}$ → $[(1/m)\, e^{(mx + c)}] + c$

Where $m \neq 0$.

Marginal and total functions

Integrate marginal functions to find total functions:

TC = ∫[d(TC)/dQ] dQ = ∫(MC)dQ

TR = ∫[d(TR)/dQ] dQ = ∫(MR)dQ

TC = Total cost;
MC = Marginal cost;
TR = Total revenue;
MR = Marginal revenue.

Matrix Algebra

A matrix

A matrix is a rectangular arrangement of numbers (elements), and an m x n order matrix has m rows and n columns. e.g. matrix A is a 2 x 2 matrix, B a 2 x 3 matrix:

A=
(1 4)
(-2 3)

B=
(0 1 1)
(2 0 1)

The transpose of a matrix is the matrix obtained by writing the rows as the columns and vice versa

For example the transpose of matrix A above:

A=
(1 -2)
(4 3)

Matrices are added by adding the corresponding elements from a row and column, but the matrices must be of the same order (e.g. both 2 x 2, or both 2 x 3) for this to be possible:

(a b) + (e f) = (a+e b+f)
(c d)   (g h)   (c+g d+h)

Matrices are subtracted by subtracting the corresponding elements from a row and column, but again the matrices must be of the same order:

(a b) – (e f) = (a–e b–f)
(c d)   (g h)   (c–g d–h)

Scalar multiplication multiplies the matrix by a scalar, an ordinary number which is constant (-2, 4, 7 etc.). With scalar k:

k (a b) = (ka kb)
  (c d)   (kc kd)

Matrix multiplication has the following rules:

(a b) x (e f) = (row 1 x **column 1**, row 1 x **column 2**)
(c d)   (g h)   (row 2 x **column 1**, row 2 x **column 2**)

$$\begin{pmatrix} a & b \\ c & d \end{pmatrix} \times \begin{pmatrix} e & f \\ g & h \end{pmatrix} = \begin{pmatrix} ae+bg & af+bh \\ ce+dg & cf+dh \end{pmatrix}$$

Matrix multiplication is only possible if the number of columns in matrix A (the first matrix) is equal to the number of rows in matrix B (second matrix).

Determinants

A determinant is a square array of numbers (elements). A 2 x 2 determinant has value:

$$\begin{vmatrix} a & b \\ c & d \end{vmatrix} = ad - bc$$

The value of a 3 x 3 determinant is given by:

$$\begin{vmatrix} a_{1,1} & a_{1,2} & a_{1,3} \\ a_{2,1} & a_{2,2} & a_{2,3} \\ a_{3,1} & a_{3,2} & a_{3,3} \end{vmatrix} = a_{1,1} \begin{vmatrix} a_{2,2} & a_{2,3} \\ a_{3,2} & a_{3,3} \end{vmatrix} - a_{1,2} \begin{vmatrix} a_{2,1} & a_{2,3} \\ a_{3,1} & a_{3,3} \end{vmatrix} + a_{1,3} \begin{vmatrix} a_{2,1} & a_{2,2} \\ a_{3,1} & a_{3,2} \end{vmatrix}$$

Where $a_{1,1}$ is the value of the number (element) in row one column one, $a_{1,2}$ the value for the element in row one, column two, etc.

Cramer's rule uses determinants to solve two simultaneous equations:

$a_1 x + b_1 y = d_1$
$a_2 x + b_2 y = d_2$

The determinants used are:

$$x = \begin{vmatrix} d_1 & b_1 \\ d_2 & b_2 \end{vmatrix} / \begin{vmatrix} a_1 & b_1 \\ a_2 & b_2 \end{vmatrix} = \Delta_x / \Delta \text{ for short}$$

$$y = \begin{vmatrix} a_1 & d_1 \\ a_2 & d_2 \end{vmatrix} / \begin{vmatrix} a_1 & b_1 \\ a_2 & b_2 \end{vmatrix} = \Delta_y / \Delta \text{ for short}$$

Where Δ_x and Δ_y differ in the two equations but Δ is the same in both.

Made in the USA
Middletown, DE
01 January 2021